Let's Go Outdoors!

LET'S GO TO A POND

By Tina Benjamin

Gareth Stevens PUBLISHING

Please visit our website, www.garethstevens.com. For a free color catalog of all our high-quality books, call toll free 1-800-542-2595 or fax 1-877-542-2596.

Library of Congress Cataloging-in-Publication Data

Benjamin, Tina.
Let's go to a pond / by Tina Benjamin.
p. cm. — (Let's go outdoors)
Includes index.
ISBN 978-1-4824-2617-5 (pbk.)
ISBN 978-1-4824-2618-2 (6 pack)
ISBN 978-1-4824-2619-9 (library binding)
1. Pond ecology — Juvenile literature. 2. Ponds — Juvenile literature. I. Benjamin, Tina. II. Title.
QH541.5 B46 2016
574.5'26322—d23

First Edition

Published in 2016 by
Gareth Stevens Publishing
111 East 14th Street, Suite 349
New York, NY 10003

Editor: Ryan Nagelhout
Designer: Nick Domiano

Photo credits: Cover, pp. 1, 5 Hero Images/Hero Images/Getty Images; p. 7 Ingrid Prats/Shutterstock.com; p. 9 Melissa King/Shutterstock.com; pp. 11, 15 oliveromg/Shutterstock.com; p. 13 Dewitt/Shutterstock.com; p. 17 SNEHIT/Shutterstock.com; p. 19 Elena Butinova/Shutterstock.com; p. 21 Oleg Golovnev/Shutterstock.com; pp. 23, 24 (geese) Noam Armonn/Shutterstock.com; p. 24 (bread) Seregam/Shutterstock.com.

Printed in the United States of America

CPSIA compliance information: Batch #CS15GS: For further information contact Gareth Stevens, New York, New York at 1-800-542-2595.

Contents

I have a pond
near my house.

Lots of animals live in the pond.

Fish swim in the water.

I go fishing
with my brother Jack.

He caught a little fish.

We throw it back!

Birds love my pond.

I see lots of ducks.

I give them bread
to eat.

Geese like to land
on the water, too!

Words to Know

bread

geese

Index